AF799891

# Call in the Air

David R. Dixon

Copyright 2013 David R. Dixon

All Rights Reserved

To Jaclyn, the one my soul loves

# Table of Contents

## A. Preflight Guide

### 1. Checkpoint

As a pilot I never had to personally face this scenario. This poem describes the decision making process of a young Marine manning a checkpoint in Afghanistan. An unknown vehicle approaches at a high rate of speed, ignoring the signs to stop for the checkpoint. Young men, very young men, such as him face extraordinarily difficult circumstances every day in combat. Trusted with binding arbitration regarding life and death, these Marines must make decisions in split seconds during confusing and chaotic circumstances.

### 2. Put on the Full Armor

Deployments are difficult mentally, physically, emotionally, and spiritually. During my second tour in Iraq, the Chaplain's office was right next door to our squadron. He was a gifted preacher, talented teacher, and just an all-around great guy. I attended his weekly studies and we developed a close friendship. When I got married in 2012 he flew to Dallas to perform the ceremony. The verses in this poem depict the passages, mostly from the Jewish Holy Scriptures, that we used to study together.

### 3. Symmetrical Stones of Grief

On 26 October 2009, my friend Captain David (Seth) Mitchell was killed while flying a combat mission in Afghanistan. I attended his funeral and since then have kept up with his father through letters and phone calls. In 2012 Seth's dad was visiting Washington DC on business, and he asked if I would join him on a trip to Arlington

National Cemetery to see Seth. That day was my 33rd birthday.

Plot 60 at ANC is where all Veterans who died after September 11, 2001 are buried. After visiting Seth's grave I continued on to see my friends Captain Kevin Kryst, Major Jeff Bland, Captain Aaron Cox, and Captain Jessica Conkling. I rarely cry, but I had a complete meltdown that day. I felt it ironic, but also appropriate, to visit a cemetery on my birthday. It was a terribly lonesome albeit cathartic experience. Captain Seth "Sniper" Mitchell is pictured flying his AH-1W Cobra over Iraq on the cover of this book (Photo credit, Porter B. Jones).

## 4. Goodnight Marines

## 5. Goodnight Navy

I don't have kids but wanted to write a series of poems for the Marine and Navy parents who are overseas and cannot be with their families at bedtime. These poems are inspired by the beloved children's book classic Goodnight Moon by Margaret Wise Brown.

## 6. The Commandants

Most Marines, me included, know little about the Commandants that served before our personal time in the Marine Corps. I went to OCS in January 2003 shortly after General Hagee took over for General Jones. This poem is meant to serve as a short familiarization about the Commandants who have boldly led our beloved Corps.

## 7. Haboob

"Haboob" is Arabic for "blasting or drafting" and is the word is used to describe the vicious sandstorms that sweep across the desert. For those not fortunate enough to be caught unexpectedly in a Haboob, one of nature's wonders, this is what it is like. Written in prose.

## 8. Alchemists

When young teenagers go off to war, the change that occurs can be profound. An Alchemist transforms lead into gold, but Marines are sometimes called to transform lead into justice - to mete out on the enemy. Written in prose.

## 9. Call in the Air

The namesake of this collection, this sonnet describes the dichotomy of warfare from the vantage point of a Forward Air Controller, calling in airstrikes from jets and helicopters. This symphony of weapons creates unimaginable chaos and disorder for the enemy.

## 10. Blood Stained Button

This poem solemnly describes a Marine's final moments with his friend. Written in prose.

## 11. Ductus Exemplo

I joined the Marine Corps three weeks after graduating from college. I had no prior military exposure whatsoever. OCS felt like 10 weeks of me constantly making mistakes and being summarily corrected by my Sergeant Instructors. It was a very tough but rewarding experience.

## 12. Flying Reminds Me

This short piece of prose relates learning to fly a Cobra to performing in a concert.

## 13. Two Funerals

Written in prose, this short poem describes the actual experience of a hero whose ship was twice sunk in World War II. Decades later my friend gave the eulogy at the second funeral for his grandfather. The story of Watertender Second Class (E-6) Jesse Paul Gilleland can be found here:

**http://www.witnesstowar.org/content/materials/Jessie%20Paul%20Gilleland.pdf**

## 14. Steps for Mankind

Fighting the war in Afghanistan or Iraq, day in and day out, consumes your life. But often, the people in America we fight to defend - they continue in their normal lives, blissfully unaware that a war is even occurring. Written in prose.

## 15. Origins Opaque

While in graduate school, I often met people who disdained America. Ironically, the military gives them the very freedom of speech that they then use to disparage the military. This poem, written in prose, describes the experience many of these students have after learning about real life in other countries. They finally come to realize that despite her problems, America is a great country, especially compared to the alternatives.

**16. Too Soon Laid to Rest**

This poem articulates the heartbreaking problem of military suicide.

**17. Belated Letter**

This short prose was inspired by the story of a Marine who was killed during the opening weeks of Operation Iraqi Freedom. His family received a letter from him three days after his death.

**18. War is Simple**

"Everything in war is simple, but the simplest thing is difficult. The difficulties accumulate and end by producing a kind of friction that is inconceivable unless one has experienced war." —Carl von Clausewitz

"At first glance, war seems a simple clash of interests. On closer examination, it reveals its complexity and takes shape as one of the most demanding and trying of human endeavors." - MCDP 1 Warfighting

**19. We Both Kneel and Nod**

In almost all wars throughout human history, each side has believed they are fighting for a noble purpose greater than themselves. Often, this "noble purpose" has a religious significance, with each side believing they are accomplishing the will of the Almighty through their belligerent efforts.

## 1. Checkpoint

Shoot or don't shoot?
Is he friendly or foe?
Taliban or Coalition,
Am I sure I really know?
God, I only finished high school, 14 months ago

Maybe he's just lost, dumb or confused.
Or is it a VB-IED?
No matter my decision
The blood will be on me.

If he's Al Qaeda and I kill him dead
a hero I'll become.
If he's just an illiterate local,
Leavenworth is my new home.

Me and all my friends are dead
if his trunk's filled with TNT.
But, if someone's going home in a pinebox
better him than me.

I believe in the cause, I swore my oath
to serve my nation well!
Liberals back home
say I'm Bush's Youth
I don't know, what the hell?

Shoot or don't shoot
I'm not sure I really know?
God I only finished high school, 14 months ago.

## 2. Put on the Full Armor

My Lord is a Warrior, He is strong and mighty to save.
He trains my hands for battle, and my heart to be brave.

My Father and fortress, friend by my side,
He subdues my enemies, in His refuge I hide.

Belt, breastplate, shoes and a shield,
helmet, a prayer, my sword I will wield.

I have put on the full armor, a mighty chariot I fly.
Will I be victorious, or will He call me to His sky?

There is a time to love and a time to hate.
Will my hands hold steady and my arrows shoot straight?

There is a time for war and for rumors of war.
Is today my last day, will I survive to see more?

We patrol the Helmand Valley of death, through the fog and shadows black.
But His rod and staff they quell my fear, and guard me from attack.

I am thrown into a lion's den, into a furnace of raging fire.
From the front I will gallantly lead my men, the timid I will inspire.

I boldly sound the righteous trumpet. My enemy I circumscribe.
After seven rounds, his walls come down, violence I cannot describe.

My adversary's blood runs crimson through the sand.
Pressed like wine from the grapes of wrath, VICTORY IS AT HAND!

Yes, my Lord is a warrior, the battle he does control
He brings tranquility to my mind, and everlasting peace to my soul.

## 3. Symmetrical Stones of Grief

Six hundred twenty four acres of sorrow,
symmetrical stones of grief

alone among friends and bitterly cold,
I stare in hollow disbelief

Some are unknown.
None are forgotten.
I ponder when this will be me
Honor, Courage, and brave Commitment,
interred in the fields of Lee

Six hundred twenty four acres of marble,
honeycombs of pain

Inconsolable,
I kneel and sob,
mourning my brothers heroically slain

## 4. Goodnight Marines

In a land overseas, in a faraway place,
there was a brave Marine, guarding his base

And behind him stood a silent platoon,
and a helicopter, flying over the moon

And there were two shiny stars and two silver bars

And a pair of dress blues and a small Navy fleet
and a pair of boots, and two yellow feet

And a flag in Iwo Jima sand
And the president's own band
and the finest dog in all the land

Now I lay in my rack, with my sleepy head down
I say Goodnight, to faithful Marines, who keep me safe and
sound

Goodnight Chesty, Goodnight Dan
Goodnight Archie, the Grand Ole Man
Goodnight Private and Corporal too
Goodnight jar and shiny shoes

Goodnight to the guards of heaven's streets
Goodnight Sergeant Major, who never retreats
Goodnight Gunny and Smokey Bear
Goodnight helicopter, floating on air

Good night Eagle Globe and Anchor
Goodnight Gunner, Goodnight Tanker
Goodnight Bronze and Silver Star
Goodnight recruits on Paris Island afar

Good night blues and scarlet stripe
Good night Barracks on 8th and I
Goodnight rifle, Goodnight flags
Goodnight sword, Goodnight dog tags

Goodnight Marine Band, softly playing Taps
Goodnight compass, Goodnight maps

Goodnight Tripoli, Belleau Wood, and Saipan
Goodnight Iraq and Afghanistan

Goodnight heroes who answer the call
Goodnight wounded Marines, and those that gave all

## 5. Goodnight Navy

On the mighty ocean, in a big grey boat,
was the United States Navy, and her crew afloat

And there were two shiny stars and two silver bars
And a uniform dress white, and a plane taking flight

And a Seabea and a bell,
And a Chief Petty Officer, with sea-stories to tell

Now I lay in my rack, with my sleepy head down
I say Goodnight to all my Shipmates, who keep me safe and sound

Goodnight Admiral, Goodnight Seamen
Goodnight Chaplain, Goodnight Midshipmen

Goodnight Chester, Goodnight Arleigh
Goodnight John Paul, and Goodnight Halsey

Goodnight Ensign flying high
Goodnight families waving goodbye
Goodnight aft, Goodnight fore
Goodnight Davey Jones on the ocean floor

Goodnight Old Ironsides, and Mighty Mo
Goodnight Arizona, in Pearl Harbor below

Goodnight Midway and Coral Sea
Goodnight SEALs, Goodnight Marines

Goodnight Bronze and Silver Star
Goodnight Corpsman in battles afar

Goodnight sweepers manning brooms
Goodnight steam and engine rooms

Goodnight Shellback and Polly-Wog,
Goodnight rain and Goodnight fog

Goodnight Port and Starboard bow
Goodnight Cooks, Goodnight chow

Goodnight hatch and port-hole round
Goodnight subs and sonar sound

Goodnight water, Goodnight bread
Goodnight torpedoes, "FULL SPEED AHEAD!"

Goodnight radar, Goodnight blip
Goodnight brave Captain "DON'T GIVE UP THE SHIP!"

Goodnight fair wind and following sea
Anchors Away the Enterprise – sail for victory

Goodnight to the gunners - who never miss their aim
Go Navy, beat Army, win the big game!

Goodnight heroes who answered the call
Goodnight wounded Sailors, and those that gave all

## 6. The Commandants

**The United States Marine Corps Commandant. A leader of the highest regard.**
**Chosen by the President, confirmed by the Senate. A warrior battle scarred.**

**They have led Marines in every clime and place where we could take a gun.**
**Now let us recognize their legacy, and call them one by one.**

Major Nicholas was the Continental Commandant
in seventeen seventy five
William Burrows started the Band,
Commandant Wharton lived at 8th and I

Anthony Gale's grave is still unknown,
the only commandant to be canned
Archibald Henderson was number five,
a wise and grand ole man

Colonel Harris was the senior Marine
in the American Civil War
General Zeilin coined the EGA,
the symbol of our Corps

Colonel McCawley chose our motto,
always faithful, Semper Fi
Charles Heywood established officer schools,
Commandant number nine

George Elliott saved the Marine Corps
from riding the Army horse
Major General Biddle
established the Advanced Base Force

George Barnett led in World War I
and landed Marines on Haiti's shore
John A. Lejeune established the hallowed tradition,
the birthday of our Corps

Major General Neville -
Marine Brevet and Medal of Honor
The Fleet Marine Force was established
by Major General Fuller

John H. Russell advanced promotions,
He landed at Vera Cruz
Thomas Holcomb, Commandant 17,
expanded the Corps for World War II

A.A. Vandegrift - the first with four stars,
he would not bend a knee
General Clifton Cates under enemy fire
led every unit of infantry

General Shepherd was on the JCS,
he earned the “French Cross of War”
Randolph Pate landed Marine battalions
on Egypt’s and Lebanon’s shores

Ike appointed Medal of Honor
recipient General David Shoup
General Green, chosen by JFK,
increased our Vietnam troops

General Chapman started the Vietnam drawdown,
Commandant under LBJ
Commandant Cushman fought in the Iwo Jima pocket,
then directed the CIA

General Wilson received the Medal of Honor,
he modernized the force
Robert Barrow acquired the Harrier,
in Korea he earned the Navy Cross

Paul Kelly was Commandant during the Beirut bombing,
he served in Vietnam
General Gray educated the Corps in Maneuver,
Commandant during Desert Storm

Commandant thirty was smart and tough,
Vietnam Vet, Carl Mundy
General Krulak created the crucible,
strategic corporal, and war blocks three

General Jones introduced the MARPATs -
Commandant during Nine Eleven
Michael Hagee was Commandant for the Baghdad March,
Operation Iraqi Freedom

General Conway led the Iraq troop surge,
MEF commander during Fallujah
Jim Amos was Commandant for Enduring Freedom,
the first aviator, OOORAH!

**One by one we have named our leader, the**
**Commandant of our beloved Corps.**
**May he continue to serve with honor**
**and lead Marines forevermore.**

## 7. Haboob

Ominous
legions marching
in random cadence. Dammit,
not again. The Heavens are completely
bisected between

blue and brown
from the sky to the ground.

A tsunami of of ravenous brunette
fruit flies. Except in some ways this is much worse.
First it's a zephyr, then a whipping gust,

SILENCE!

gale force winds furiously driving sand into
every crevice of my body. I am enveloped
in a brown haze. I inhale this brown baby powder
into my mouth and then remember, that the
sand in this godforsaken country
has a high fecal content. I am nauseas
and want to spit,
but,
then I would have to open my mouth.

Millions of pellets, a blurry, brown,
cat-of-nine-tails whip by face.

This t-shirt around my mouth is worthless.

Screw it, laugh, what else can I do?
This Haboob will pass
in a day or two.

## 8. Alchemists

Alchemists transform lead into gold

gold - wealth and joy

wealth creates and sustains
a prosperous abundant life.

Alchemists transform lead into life
life, with infinite value and meaning

but life
this life we live
is quickly quelled
by a conical shard of lead
weighing no more than 9 grams.

M-16A2, .556 ball and tracer

fear,
thrill,
survival,
confusion

Alchemists transform lead into life
I am 19, and I kind of do

the opposite.

Yet I am the one transformed.

## 9. Call in the Air

as a Forward Air Controller you are sometimes
the audience but often the composer of this symphony of
slaughter:

**a musical fluidity**
**of harrowing violence**
**a dizzying rapidity**
**of heroic valiance**

**a surgical precision**
**of savage atrocity**
**undiscerning bludgeon**
**and selfless generosity**

**clockwork killing**
**my conscience confounds**
**wild-eyed thrilling**
**disorder abounds**

**To the Constitution, oaths we swore**
**Call in the Air - This is War!**

## 10. Blood Stained Button

a green nickel

four identical holes

nothing imprinted
no Monticello home
no god to trust

only an olive thread and a sticky stain
reminds me of maple syrup -

waffles at my granddad's summer house in the Cape
you never forget that smell

flesh seared by incendiary steel

the grenade was just doing its job
but I failed to do mine

I am hypnotized by this damn button

the rancid odor

I taste death in my portals

I want to tell my buddy that I love him
and that it will be ok
and we will be fishing in the Adirondacks soon

but in the last 90 seconds that I had with him
all I could do was gape entranced at this
button on charred camouflage
loosely draped over the shredded
remains of his once Adonis figure.

Love
War

It is not all fair

## 11. Ductus Exemplo

Lost is my normal course,
a Candidate leading a squad.
Common sense is **not** a common virtue,
my every decision is flawed.

Sergeant Instructors surround me,
wolves upon their prey.
LOUDER FASTER HARDER BETTER
My stupidity on display.

## 12. Flying Reminds Me....

...of a piano recital

performing at a sold-out concert

The Kennedy Center

ANTICIPATION

hands
feet
eyes
ears

all working in seamless harmony

masterfully manipulating rudders and wings

keys and strings

A crescendo of noise - the sound of freedom

ANXIETY

the brain thinks ahead

muscle memory
trust your instrument

EXHILARATING

playing solo for the first time

flying solo for the first time

much more difficult

both intoxicating
and addictive

## 13. Two Funerals

they held a funeral
in honor of my grandfather

he was not there
Missing in Action

how long can you tread water?
are you scared of sharks?

then he showed up.
my grandfather
on his mother's doorstep
(my great grandmother)
I guess it was about
9 months later

she screamed and fainted
as if he was a ghost
which in a way, he was

risen from the graveyard
of a POW camp

how would you react?

it wasn't his fault
his family lived in rural Georgia
and the phones and telegrams,
as I understand,
were not all that reliable
back when we were fighting the Japanese

I gave his eulogy
fifty seven years later
I was honored to honor my hero

how many sailors do you know
who have had two funerals?

## 14. Steps for Mankind

everyone paid attention
When Neal stepped on
that boulder of dust
and planted the flag on
that barren, undiscovered wasteland

We walked through a
desert for twelve months

sand, more like,
moon dust -
brown baby power

an effluvia of war,
sacrifice, and patriotism
wisped away in a
cloud each time
our soles touched down

but nobody noticed
our steps for mankind

we went to war
and America went

to the Galleria

## 15. Origins Opaque

in college I was assigned a random roommate
a foreigner
he received political asylum to come here

**(his dad was a doctor)**

his country has been engaged
in a vicious border war, for maybe a decade
with a militaristic neighbor to the north;

an influx of destitute refugees from the nation to the south
is overwhelming their fractious services and infrastructure.

a drought and backwards agricultural practices
have further exacerbated this deplorable situation

**(and neighbors with the ambassador)**

the origins of the border conflict - opaque
even to warlords who command the carnage
and the child-soldiers who execute orders

but yet it continues unabated - this war in his homeland

looting, subversion, sabotage, terrorism
best describe his totally corrupt government

over the semester, as he told me stories
from his childhood, I thought to myself

"Maybe I should put my hand over my heart
before the football games, instead of just standing there

indignant"

## 16. Too Soon Laid to Rest

he left a teenager and returned full grown
no one understands, now he battles alone

reflect without judgment this issue outstanding
a condition replete with misunderstanding

one night he is drinking and smoking cigars
the next draped in cloth, be-speckled with stars

a hero and warrior too soon laid to rest
with confusion and sadness long suppressed

self-inflicted wounds may seem enigmatic
he bears invisible scars of stress traumatic

now ponder America this problem unique
safe back from war, but dead the next week

## 17. Belated Letter

My older brother Sebastian was killed
in a helicopter crash overseas

we got the call on a Monday
I was still drunk on Thursday
when our mother
received a postcard from him

Sebastian died in battle 72 hours ago

but the mail takes 10 to 14 days
from the other side of the world

mom sobbed despondently
when she read his letter

"of course"

I thought to my self

"Sebastian was always late for everything."

## 18. War is Simple

My friends back home,
And people I meet at bars,

These people,

they often ask me,
"What was it like?"

well...it's simple.

it's us versus them, good versus evil
a narrative old as time
and good always wins, right?

so, war seems simple
simple violence that is,
the main ingredient to this recipe of ruin, where
adrenaline marinates in numbing fatigue

do you honestly want to know what it's like?

Listen.

love for my buddies coalesces
with abhorrence toward my enemy
and my apathy for politics.

a quagmire.

that devolves into a carrousel of carnage
humming around a masterpiece of destruction.

this is what it is like

this...cinema of slaughter
where teenagers govern sophisticated
weapons with adolescent emotion

this...this...opera of bedlam,
where pride for your nation
collides in cacophony with the sobering humility
of life's fragility

war

at first,
it seems simple in nature

but then,
it is impossible to explain

So I don’t

## 19. We Both Kneel and Nod

ordained by "god" you piously boast.
led by his holy heavenly host?

with blood stained hands you fast applaud.
and venerate the "works of god"

now sit and answer this question anew
which "god" in fact approves of you?

morally corrupt and brutally wise
you bade your fighters terrorize

crimes too sordid for printer's pages
scarcely matched in barbarous ages

the women you abuse and cruelly deprive
the children you burned and buried alive

churches reduced to cinders and dust
rapes and tortures with bestial lust

insufficient parchment and quill exist
to account the wretched lies you twist

but now I ponder with curious mind
that **YOU** accuse **ME** of evil in kind

we each believe we're justly condoned
by Him who judges divinely enthroned

but when this war's over, we'll both kneel and nod
and finally learn who was guided by God

* *Inspired by The Kaiser and God by Barry Pain*

**The views and ideas in this book are my own and do not reflect the official position of the Marine Corps or Department of Defense in any way.**

Made in the USA
Monee, IL
04 November 2026

67387901R00020